Searchlight BOOKS™

Spotlight on Climate Change

Climate Change Activism

Tracy Sue Walker

Lerner Publications ◆ Minneapolis

For Mom and Dad

Lerner Publications Company
An imprint of Lerner Publishing Group, Inc.
241 First Avenue North
Minneapolis, MN 55401 USA

For reading levels and more information, look up this title
at www.lernerbooks.com.

Main body text set in Adrianna Regular.
Typeface provided by Chank.

Editor: Brianna Kaiser
Lerner team: Sue Marquis

Library of Congress Cataloging-in-Publication Data

Names: Walker, Tracy Sue, author.
Title: Climate change activism / Tracy Sue Walker.
Description: Minneapolis : Lerner Publications, [2023] | Series: Searchlight books
    - spotlight on climate change | Audience: Ages 8–11 | Audience: Grades 4–6 |
    Summary: "Recent years have seen an increase in discussions about climate
    change. But what are people doing about it? Discover the important people and
    movements that are leading the fight against the climate crisis"— Provided by
    publisher.
Identifiers: LCCN 2021045340 (print) | LCCN 2021045341
    (ebook) | ISBN 9781728457901 (library binding) | ISBN 9781728463919 (paperback) |
    ISBN 9781728461892 (ebook)
Subjects: LCSH: Climatic changes—Juvenile literature. | Environmentalism—Juvenile
    literature.
Classification: LCC QC903.15 .W37 2023 (print) | LCC QC903.15 (ebook) | DDC
    320.58—dc23/eng/20211119

LC record available at https://lccn.loc.gov/2021045340
LC ebook record available at https://lccn.loc.gov/2021045341

Manufactured in the United States of America
1-50818-50157-12/2/2021

Table of Contents

Chapter 1

A PLANET IN CRISIS

Melting polar ice caps, rising sea levels, and raging wildfires. Why is Earth facing these problems? The answer is climate change. Human activities such as driving cars trap heat in the atmosphere and cause Earth's temperature to rise quickly. More heat in the atmosphere makes natural disasters, such as floods and droughts, more likely.

The hottest decade ever recorded was between 2010 and 2019. In Death Valley, California, two of the hottest temperatures ever recorded on Earth were in 2020 and 2021. Temperatures reached around 130°F (54°C) both years. Concerned people are taking action. Climate change activists are fighting for our planet.

Climate change can make dangerous events such as wildfires more common.

Rise of the Climate Change Movement

Scientists grew concerned about climate change in the 1950s and 1960s. People around the world began paying more attention to the problem. On April 22, 1970, the first Earth Day was held. The event helped raise awareness about the warming planet. In the 1970s the US government passed laws to help protect the environment.

The United Nations formed a climate change panel in 1988. The panel reviews the science on climate change. It also gives information on climate change to governments.

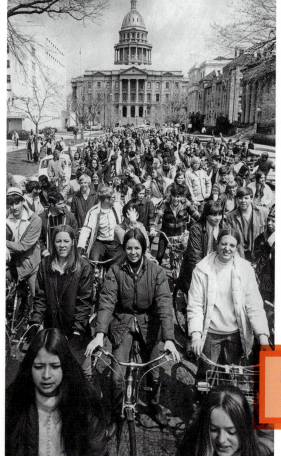

People taking part in Earth Day in 1970

More people started looking at how burning fossil fuels causes global warming. Fossil fuels formed underground in Earth millions of years ago. They are used to produce energy. Some commonly used fossil fuels are oil, coal, and natural gas.

Larger numbers of people began holding climate change protests in the early 2000s. These protests continue. And more than one billion people worldwide participate in Earth Day.

Lisa Sipaia-Baker of the group Pacific Climate Warriors speaks to a crowd in 2021. The group works to protect the Pacific Islands from climate change.

Spreading the Word

Activism involves sharing information and getting others involved. Activists do this in many ways. They often talk to community members and ask government leaders to support climate change laws. They also hold protests and strikes.

Some activists speak on radio and news shows or write books and articles. Others take pictures and make films. Activists also reach many people through social media.

Climate change activism is more important than ever. Earth's climate is changing quickly. And activism can lead to a decrease in carbon footprints.

Vanessa Nakate, a climate change activist from Uganda, speaks at a rally in 2021.

STEM Spotlight

Carbon footprints are the amount of greenhouse gas created by humans. Transportation, housing, and food add to a person's carbon footprint. Greenhouse gases, such as carbon dioxide, cause global warming by trapping heat in Earth's atmosphere.

Many businesses and governments are taking steps to reduce their carbon footprints. For example, the entertainment industry is looking into using solar energy for power and reusable materials for building sets. Carpooling and turning off unused lights are ways to reduce your carbon footprint.

Chapter 2

A GLOBAL IMPACT

People are joining forces to fight against Earth's climate crisis. Their actions are having an impact on businesses and governments around the world.

Many people join global climate movements and work to end the use of fossil fuels. Members of the movements want to turn to renewable energy. They also want to help communities that are the hardest hit by climate change. They have peaceful protests, give speeches, and hold community cleanup days.

United Nations Conferences on Climate Change

In June 1992, the United Nations Earth Summit was held in Rio de Janeiro, Brazil. At the conference, many countries agreed to help control greenhouse gases and signed a treaty.

STUDENTS AT THE 1992 EARTH SUMMIT

Former president Barack Obama (*right*) meets with other world leaders at the 2015 UN Climate Change Conference.

The Kyoto Protocol treaty was adopted in 1997 in Kyoto, Japan. It extended the treaty from the Earth Summit. Countries agreed to reduce greenhouse gas emissions. Under the treaty, countries' emissions were watched and recorded. The first commitment period of the Kyoto Protocol ended in 2012.

The 2015 UN Climate Change Conference was held in Paris, France. Almost two hundred countries adopted the Paris Agreement. The goal of this treaty is to limit global warming and achieve zero greenhouse gas emissions by mid-century. In 2017, the US became the first country to leave the Paris Agreement. The US rejoined in January 2021.

President Joe Biden gives a speech about climate change in January 2021.

A great deal still must be done to achieve the agreement's goals. But more countries and companies are now establishing targets of low and zero greenhouse gas emissions.

PEOPLE ALL OVER THE WORLD PARTICIPATE IN CLIMATE STRIKES, SUCH AS THIS 2019 STRIKE IN THAILAND.

STEM Spotlight

Renewable energy is made from energy sources that are replaced by nature and do not pollute the environment. Sustainable energy comes from sources that can never be used up. Some energy sources are both renewable and sustainable. These include wind, solar (sun), and geothermal (heat inside Earth).

Renewable sources are less harmful for the environment than nonrenewable sources, such as coal. They create energy without using fossil fuels, which produce greenhouse gas emissions. Using renewable energy sources can greatly reduce carbon footprints.

The Climate Pledge

The Climate Pledge was created in 2019. Its goal is for companies to achieve zero carbon emissions by 2040. When groups take the pledge, they agree to regularly report greenhouse gas emissions and reduce carbon emissions to reach the goals of the Paris Agreement. Over one hundred of the world's leading businesses have signed the pledge.

At the 2021 Youth4Climate event in Milan, Italy, hundreds of young people helped create strategies to address climate change.

Chapter 3

YOUTH CLIMATE CHANGE ACTIVISM

Millions of young people around the world are fighting climate change. They are forming strikes and taking leadership positions in groups. Social media plays a large role in youth-led activism. It allows young people around the globe to connect easily over a common cause.

Greta Thunberg speaks at a Fridays for Future event in 2021.

Since Swedish activist Greta Thunberg's first climate strike in 2018, youth in over one hundred countries have started striking. They are challenging government and business leaders that are not taking action against climate change or deny the crisis.

Youth Leadership

The Youth Adaptation Network is part of the Global Center on Adaptation (GCA). GCA is one of the world's leading groups on climate change adaptation. Climate change adaptation happens when a country or people adjust to the current effects of climate change and prepare for changes in the future.

Young people are given the chance to learn from leading climate change professionals at GCA offices. These young leaders are then able to share their knowledge on climate change with their communities.

GCA OFFICE IN ROTTERDAM, NETHERLANDS

The Sunrise Movement was founded in 2017. Its mission is to stop climate change and create jobs. Its members are young people who share climate change information with their communities. The Sunrise Movement has over four hundred locations across the US.

On June 28, 2021, climate change activists participate in a march organized by the Sunrise Movement.

The United Nations started the Youth Advisory Group on Climate Change in 2020. Young climate change leaders are chosen from around the world to participate. They assist in the UN's decision-making and strategy.

Xiuhtezcatl Martinez

Activist Xiuhtezcatl Martinez has inspired youth with his participation and leadership in movements for climate change and social justice.

In 2017, he addressed the United Nations on climate change. He was a youth director at Earth Guardians until 2019. Earth Guardians teaches youth to be activists.

Martinez is also a musician. His music expresses his thoughts on the climate change crisis and other issues, including the impact of fossil fuels on Indigenous communities.

Xiuhtezcatl Martinez in 2017

Chapter 4

LEADING THE WAY

Many people are leading the way in climate change activism. Their leadership motivates people and governments to act against climate change.

Inspiring Change

Greta Thunberg was born in Sweden in 2003. When she was in third grade, she saw films about global warming that sparked her interest in climate change. Years later, she was inspired by American students holding strikes in support of gun reform after the school shooting in Parkland, Florida, in February 2018.

On August 20, 2018, Thunberg sat outside the Swedish Parliament House with a sign that said School Strike for Climate in Swedish. She called on the Swedish parliament to take more action against climate change.

Thunberg's strike inspired Fridays for Future. In the movement, students skip school on Fridays to bring attention to climate change. On March 15, 2019, over one million people in more than one hundred countries went on strike to demand action against climate change.

Thunberg sits outside the Swedish Parliament House in 2018.

Rhiana Gunn-Wright

Rhiana Gunn-Wright grew up on Chicago's South Side. She saw how low-income communities and communities of color are more negatively impacted by climate change than higher-income communities. This inspired her career in working to help the environment and fighting for social justice.

Gunn-Wright helped develop the Green New Deal. It called for the federal government to reduce greenhouse gas emissions, move away from using fossil fuels, and invest in clean energy.

Rhiana Gunn-Wright (*second from right*) poses with others at the 2019 Town & Country Philanthropy Summit.

Anthony Nyong is one of the world's leaders in working to protect the environment. He was born in Nigeria, West Africa. One of his goals is to help Africa develop tools to combat climate change.

Nyong is the regional director for Global Center on Adaptation (GCA) Africa. GCA Africa formed in 2020 and helps protect communities in Africa from the effects of climate change. In 2019, Apolitical magazine named Nyong in the top twenty of their list of the world's most influential people in climate policy.

Vandana Shiva is a scientist, activist, and author. She believes ending world hunger is possible and fights for farming that is better for the environment. She founded Navdanya in 1991. Navdanya is a movement to protect living resources like native seeds. In 2003, *Time* magazine named her an environmental hero.

Vandana Shiva
in 2017

Taking Action

Climate change is happening. But people all over the world are taking action. By getting informed and spreading the word, you can help too.

People hold signs in a climate march in London, England, in 2015.

You Can Help!

You can be a climate change activist. A great way to make a difference is by writing to your elected officials about climate change. Here are some websites and a sample letter to help you get started:

Helpful Websites

- Open States, https://openstates.org
- United States House of Representatives, https://www.house.gov/representatives
- United States Senate, https://www.senate.gov/senators/senators-contact.htm

Sample Letter

Dear [ELECTED OFFICIAL'S NAME]:

My name is [NAME], and I am a [GRADE LEVEL] student in [TOWN]. I am writing because I am very concerned about the climate change crisis. This issue is important to me because [LIST TWO OR THREE REASONS AND HOW THEY AFFECT YOU].

I am asking you to please help in finding a solution for our community and planet. I hope you will consider bringing this issue to [STATE LEGISLATURE/CONGRESS, etc.].

Thank you so much for your time and consideration. I hope together we can truly make a difference.

Sincerely,
[NAME]

Glossary

activist: a person who uses or supports strong actions to stand up for a cause

adaptation: changing or adjusting

climate change: a change in the usual weather conditions of a place over a long time

emission: a substance released or sent out into the atmosphere

environment: the things and conditions around a person

fossil fuel: a fuel containing carbon that is formed from prehistoric animal and plant remains

greenhouse gas: a gas in Earth's atmosphere that traps heat from the sun

policy: a plan or rules that are used to guide actions in a government or organization

strike: when a person or group of people refuses to continue working to force an employer or organization to make a change

treaty: an agreement between two or more countries or nations

Learn More

Doeden, Matt. *Greta Thunberg: Climate Crisis Activist*. Minneapolis: Lerner Publications, 2021.

Kids against Climate Change
https://kidsagainstclimatechange.co

Klein, Naomi. *How to Change Everything: The Young Human's Guide to Protecting the Planet and Each Other*. New York: Atheneum Books for Young Readers, 2021.

NASA Climate Kids: A Guide to Climate Change for Kids
https://climatekids.nasa.gov/kids-guide-to-climate-change/

National Geographic Kids: Climate Change
https://kids.nationalgeographic.com/science/article/climate-change

Thunberg, Greta. *No One Is Too Small to Make a Difference*. New York: Penguin Books, 2019.

Index

Photo Acknowledgments

Image credits: Go Nakamura/Bloomberg Creative/Getty Images, p. 5; Duane Howell/The Denver Post/Getty Images, p. 6; Joshua Prieto/SOPA Images/Shutterstock.com, p. 7; Stefano Nicoli/Speed Media/Shutterstock.com, p. 8; Antonio Ribeiro/Gamma-Rapho/Getty Images, p. 11; David Silpa/UPI/Shutterstock.com, p. 12; Anna Moneymaker/POOL/EPA-EFE/Shutterstock.com, p. 13; Narong Sangnak/EPA-EFE/Shutterstock.com, p. 14; Quirinale Press Office/AGF/Shutterstock.com, p. 16; Mauro Ujetto/NurPhoto/Shutterstock.com, p. 18; Shutterstock.com, p. 19; Chip Somodevilla/Getty Images, pp. 20–21; Cindy Barrymore/Shutterstock.com, p. 22; Michael Campanella/Getty Images, p. 24; Bryan Bedder/Stringer/Getty Images, p. 25; Kevin Kane/Stringer/Getty Images, p. 27; James Gourley/Shutterstock.com, p. 28.

Cover: AP Photo/Agencia El Universal/Lucía Godínez/JMA.